YOUR KNOWLEDGE HAS VALUE

- We will publish your bachelor's and
  master's thesis, essays and papers

- Your own eBook and book -
  sold worldwide in all relevant shops

- Earn money with each sale

Upload your text at www.GRIN.com
and publish for free

Implementing a Synthetic Smart Grid Design using Matlab

Mirroyal Ismayilov

Bibliographic information published by the German National Library:

The German National Library lists this publication in the National Bibliography; detailed bibliographic data are available on the Internet at http://dnb.dnb.de.

ISBN: 9783346865601
This book is also available as an ebook.

FEMTO-ST Institute

International Master GREEM
Automation and Robotics
Control for Green Mechatronics

SMART GRID DESIGN WITH MATLAB

Author :
MIRROYAL ISMAYILOV

Report
22-01-2022

Contents

List of Figures

List of Tables

Listings

1 Introduction

1.1 Smart Grid

The Smart Grid represents an unprecedented opportunity to move the energy industry into a new era of reliability, availability, and efficiency that will contribute to our economic and environmental health. During the transition period, it will be critical to carry out testing, technology improvements, consumer education, development of standards and regulations, and information sharing between projects to ensure that the benefits we envision from the Smart Grid become a reality. The benefits associated with the Smart Grid include:

- More efficient transmission of electricity

- Quicker restoration of electricity after power disturbances

- Reduced operations and management costs for utilities, and ultimately lower power costs for consumers

- Reduced peak demand, which will also help lower electricity rates

- Increased integration of large-scale renewable energy systems

- Better integration of customer-owner power generation systems, including renewable energy systems

- Improved security

Table 1: Smart Grid Elements

Power Sources	Storage	Loads
Solar photovoltaic (PV)	Flywheels	Daily profiles with seasonal variation
Wind turbine	Customizable batteries	Deferrable (water pumping, refrigeration)
Generator: diesel	Flow batteries	Thermal (space heating, crop drying)
Electric utility grid	Hydrogen	Efficiency measures
Traditional hydro		
Run-of-river hydro power		
Biomass power		
Generator		
Microturbine		
Fuel cell		

1.2 Smart Grid Design with MATLAB

In this practical work, a synthetic smart grid design is implemented by using MATLAB. The specific focus is on the way the smart grid design problem has been turned into an optimization problem. Cost function is defined and minimized by the help of Global Optimization Toolbox, specifically, by the help of Genetic Algorithm within MATLAB.

Genetic Algorithm
Genetic algorithm solves smooth or nonsmooth optimization problems with any types of constraints, including integer constraints. It is a stochastic, population-based algorithm that searches randomly by mutation and crossover among population members.

2 Selected Location and Load Profile

The selected location is considered to be **Rural African Region (Botswana)**.

Editor's note: the figure was removed due to copyright issues.

Figure 1: Rural Africa, Botswana

Load profile is approximated from an **Household Load** (Resource: NREL Data Catalog) with the specification shown below:

SUMMARY OF RURAL AFRICAN HOUSEHOLD LOAD PROFILES (W)

	High Income Household	Medium Income Household	Low Income Household	All Households Assuming Varying % High, Medium, and Low
0:00	8.5	2.0	0.5	3.7
1:00	8.5	2.0	0.5	3.7
2:00	8.5	2.0	0.5	3.7
3:00	8.5	2.0	0.5	3.7
4:00	8.5	2.0	0.5	3.7
5:00	8.5	2.0	0.5	3.7
6:00	10.0	2.9	0.8	4.6
7:00	10.0	2.9	0.8	4.6
8:00	29.9	8.9	1.9	13.6
9:00	29.9	8.9	1.9	13.6
10:00	29.9	8.9	1.9	13.6
11:00	32.1	10.3	2.4	14.9
12:00	18.0	5.3	1.5	8.3
13:00	18.0	5.3	1.5	8.3
14:00	18.0	5.3	1.5	8.3
15:00	11.4	2.6	0.6	4.8
16:00	11.4	2.6	0.6	4.8
17:00	28.4	13.2	4.5	15.3
18:00	42.6	21.1	8.1	23.9
19:00	75.9	40.3	16.3	44.1
20:00	112.8	60.5	23.5	65.6
21:00	107.4	56.1	21.4	61.6
22:00	55.4	28.0	11.2	31.5
23:00	15.3	6.5	2.8	8.2
Total Wh/day/househo|	707.0	301.7	106.5	371.7
Total kWh/year/househ	258	110	39	136

Figure 2: Rural African Household Load Profiles

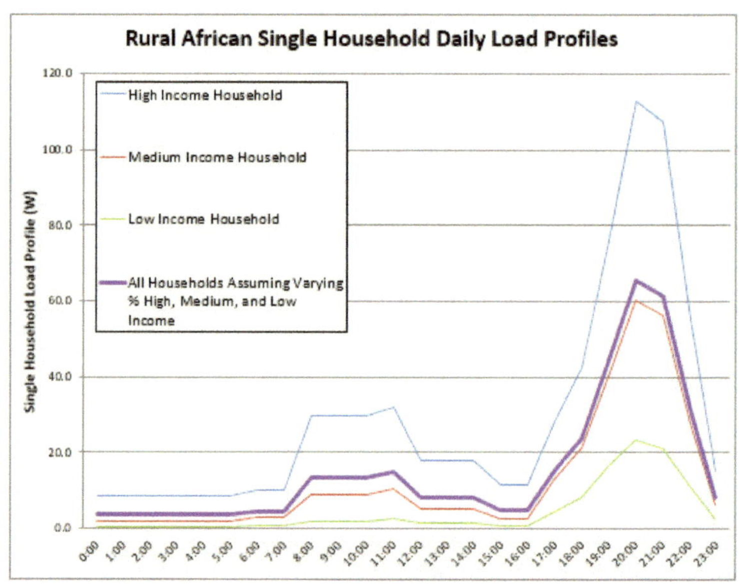

Figure 3: Rural African Household Load Profiles Graphical Representation

Here, the last colomn values (**All Households Assuming Varying % High, Medium, and Low**) are taken as an input to our Load profile which is approximated in Matlab. After last coloumn is taken as an input, it is replicated and multiplied with 300 in order to generate a load for **300 houses**.

Listing 1: Generation of Hourly Load profile for a year with Matlab

```
clc
clear
close all
%% Generation of Hourly Load profile for a year with Matlab
load 'LOAD.csv'
LD=300*repmat(LOAD,1,365); %Replication of an Load array
Load=LD+0.3*rand(size(LD)).*LD-0.3*rand(size(LD)).*LD; %
    Randomizing the loads for the day
MLDR=mean(Load,1); % Mean load of each day
TD=1:8760; % Indexing hours of the year
DY=1:365; % Indexing days of the year
figure (1)
plot(TD,Load(:),'r');
hold on
xlabel('Hour of Year');
ylabel('Load Profile for 300 houses [W]');
```

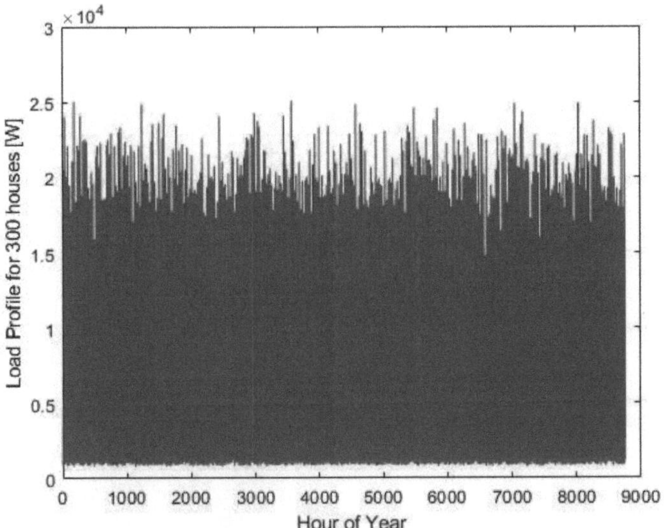

Figure 4: Seasonal Load profile for 300 houses

3 Renewable Energy Resources

Renewable energy data is also randomized data which is initially taken from HOMER Pro based on the location we specified earlier. The following code shows how the data is randomized.

Listing 2: Generation of Hourly Renewable Energy Data for duration of a year with Matlab

```matlab
%% Generation of Hourly Renewable Energy Data
load 'WIND_S.csv';load 'SOLAR_R.csv'
Ws=repmat(WIND_S,1,24*31);
Ws=Ws';Ws=Ws(:);Ws(8761:end)=[];
Ws=Ws+0.3*rand(size(Ws)).*Ws-0.3*rand(size(Ws)).*Ws;    % Wind
    Speed
figure (2)
plot(TD,Ws(:),'b');
hold on
xlabel('Hour of Year');
ylabel('Wind Speed [m/s]');
Sr=repmat(SOLAR_R,1,24*31);
Sr=Sr';Sr=1000*Sr(:)/24;Sr(8761:end)=[];
Sr=Sr+0.3*rand(size(Sr)).*Sr-0.3*rand(size(Sr)).*Sr; % Solar
    Radiation
figure (3)
plot(TD,Sr(:),'k');
```

```
16  hold on
17  xlabel('Hour of Year');
18  ylabel('Solar Radiation [W/m^2]');
```

3.1 The Wind Resource

The hourly wind speed data is plotted below for a duration of 8760 hours (duration of a year) for Botswana. As it can be observed from the graph, the wind speed is higher in the second half of the year around August to October. The lowest wind speed data is observed around April and May.

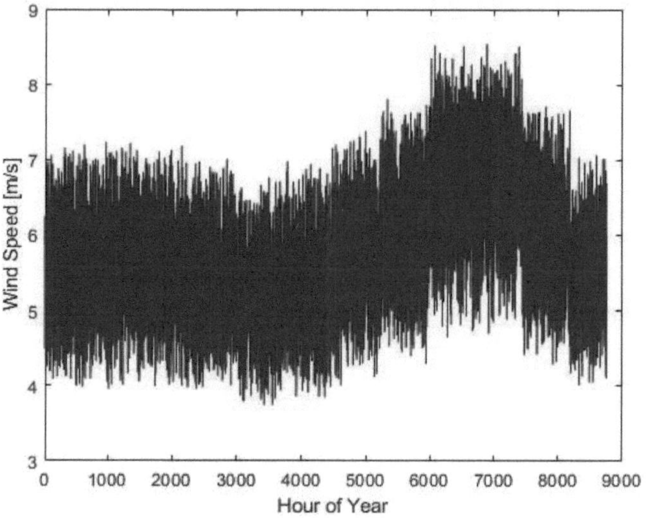

Figure 5: Hourly Wind Speed data for duration of a year

3.2 The Solar GHI Resource

The hourly solar radiation data is plotted below for a duration of 8760 hours (duration of a year) for Botswana.
As it can be seen from the graph, the solar radiation is higher in the beginning and the end of the year. The lowest solar radiation data is observed around May and June. The data makes sense since the data is taken from the Southern Hemisphere (Botswana) where the winter is between May and September.

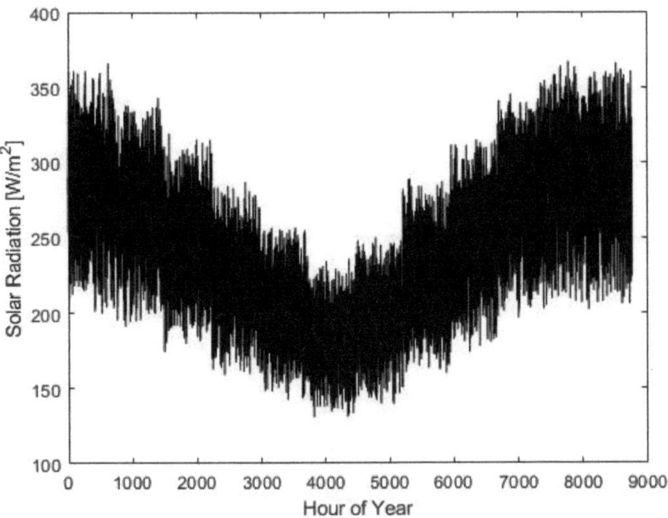

Figure 6: Hourly Solar Radiation data for duration of a year

4 Component Selection and Pricing

First of all, initial schematic design diagram should be built just in order to figure out the optimization variables which in this case will be generator size, wind turbine quantity as well as quantity of photovoltaic and battery modules.

Normally, Smart Grid Design has following structure:

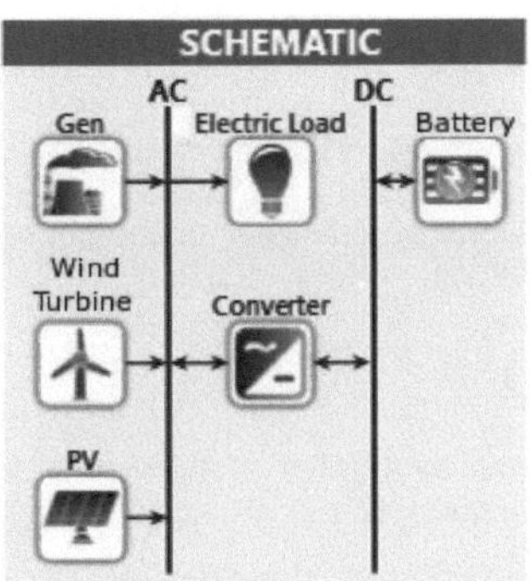

Figure 7: Schematic Design Diagram

4.1 Photovoltaic Panel

The selected photovoltaic panel is **Solar panel 110W 12V Mono - SuperWatt**. The properties of the photovoltaic panel is illustrated below:

ELECTRICAL SPECIFICATIONS

Panel type	12V panel	Power Pmax	110W
Short-circuit current* Isc [A]	5.66	Open circuit voltage* Uoc [V]	22.65
Current at MPP* Impp [A]	5.29	Voltage at MPP* Umpp [V]	18.9
Efficiency η [A]	15.15	Measurement tolerances	+/-3%

Figure 8: Solar panel 110W 12V Mono - SuperWatt

Price of the selected module: **64.39 €/piece**

With the current exchange rate:

Price of the selected module: **73.07 \$/piece**

Power calculation for a single photovoltaic panel is expressed below:

$$P_s = R_s \cdot \eta_{PV} \cdot A_{PV} \tag{1}$$

where R_s, η_{PV}, A_{PV} are solar radiation, efficiency of the photovoltaic component, surface area of the photovoltaic panel, respectively.

As it is shown in Fig. 8, η_{PV} is 15.15.

For simplcity, surface area of the panel A_{PV} is considered to be 1 m^2. So, the generated solar power P_s is varying with respect to the amount of solar radiation which is indicated as R_s.

4.2   Wind Turbine

The selected wind turbine unit is **AWS HC 1.8 kW**. The properties of the horizontal turbine unit is illustrated below:

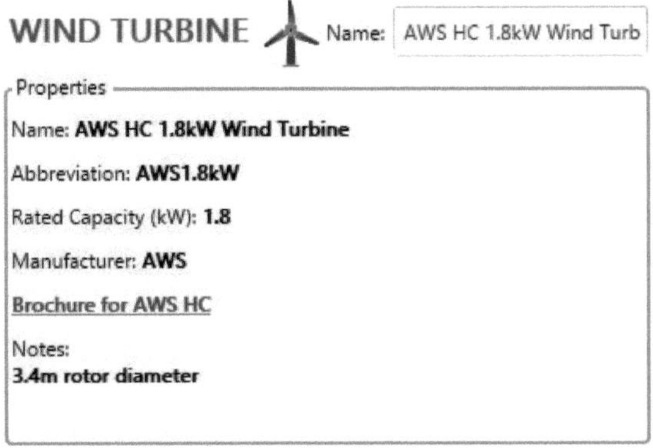

Figure 9: AWS HC 1.8 kW

Price of the selected module: **4800 \$/piece**

Operational/Maintenance cost for **a single component** is **20 \$/year**
In order to convert the wind speed to energy, the power curve for the selected wind turbine component must be analyzed.

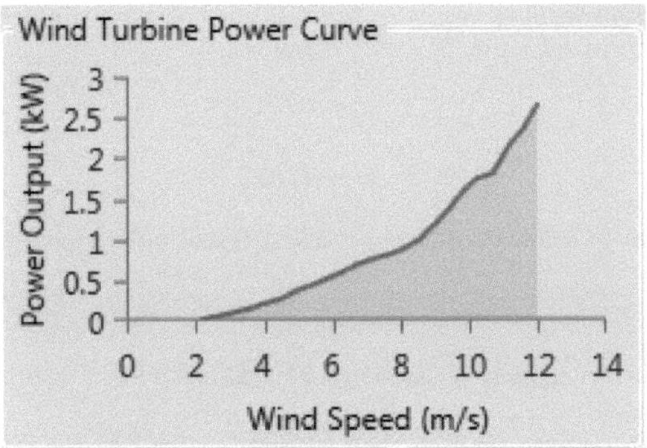

Figure 10: AWS HC 1.8 kW Power Curve

Since it is better to simplify the power curve for wind turbine. It is proposed to make an equivalent power curve for selected wind turbine.

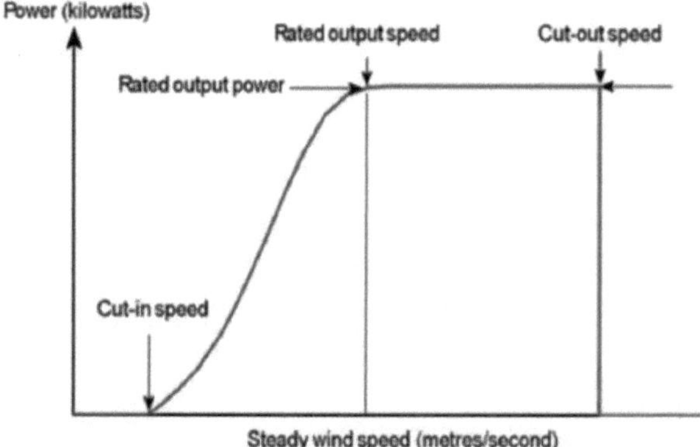

Figure 11: Simplified Equivalent Power Curve

Power calculation for the equivalent curve is determined by the equation below:

$$P_w = \begin{cases} 0 & \text{if } s_w < s_{ci} \\ P_r \cdot \frac{(s_w - s_{ci})}{(s_{co} - s_{ci})} & \text{if } s_{ci} \leq s_w \leq s_{co} \\ P_r & \text{if } s_w > s_{co} \end{cases} \tag{2}$$

The equations simply means that when the wind speed is less than Cut-in speed, no power is generated by wind turbine. When wind speed is between the Cut-in and Cut-out speed values, generated power is based on the gradient of the power curve shown above. Lastly, if the wind speed is more than the Cut-out speed, generated power is assumed to be equal to the Rated output power of the wind turbine.

The parameters for equivalent curve is shown below:

$$s_{ci} = 4 \left[\tfrac{m}{s}\right]$$
$$s_{co} = 10.7 \left[\tfrac{m}{s}\right]$$
$$P_r = 1800 \left[W\right]$$

The s_{ci}, s_{co}, P_r are Cut-in speed, Cut-out speed and Rated output power, respectively.

4.3 Battery

The selected kinetic battery model is **Discover 12VRE-3000TF**. The properties of the battery model is illustrated below:

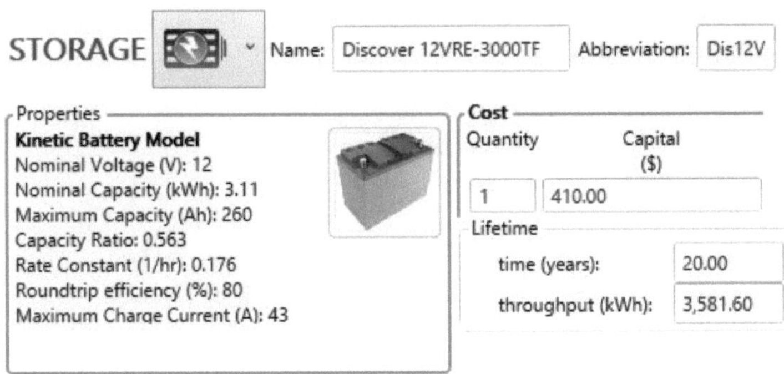

Figure 12: Discover 12VRE-3000TF

Price of the selected module: **410 \$/piece**

4.4 Generator

The selected stationary generator model is **Generac 20kW Protector**. The properties of the generator model is illustrated below:

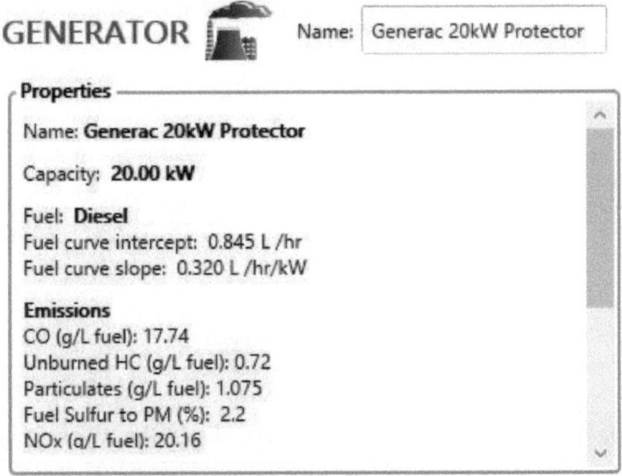

Figure 13: Generac 20kW Protector

Price of the selected module (Initial Capital Cost): **14400 \$/piece**
In this practical work, it is assumed that single generator unit is utilized.

Energy Unit Cost

In Botswana, average cost for diesel per gallon is around \$4.077. A 20-kilowatt generator will use up about 1.6 gallons of fuel per hour while running at full load. A ballpark cost would be

$156.6 for a day of run time as the cost to run the generator. The energy cost for single energy unit can be calculated as below:

$$24\,[hrs] \rightarrow 156.6\,[\$] \rightarrow 20\,[kW]$$
$$1\,[hr] \rightarrow 6.525\,[\$] \rightarrow 20\,[kW]$$
$$1\,[hr] \rightarrow 0.326\,[\$] \rightarrow 1\,[kW]$$
$$1\,[hr] \rightarrow 0.326 \cdot 10^{-3}\,[\$] \rightarrow 1\,[W]$$

Maintenance Cost

Since for the energy unit cost, we considered that generator is assumed to be at full load, maintenance cost is forfeit due to this safety which is mentioned above. Energy cost unit is powerful enough, so that maintenance cost is not considered.

5 Smart Grid Design

The smart grid design is implemented as an algorithm in Matlab, the structure of the algorithm as a flow chart is illustrated below:

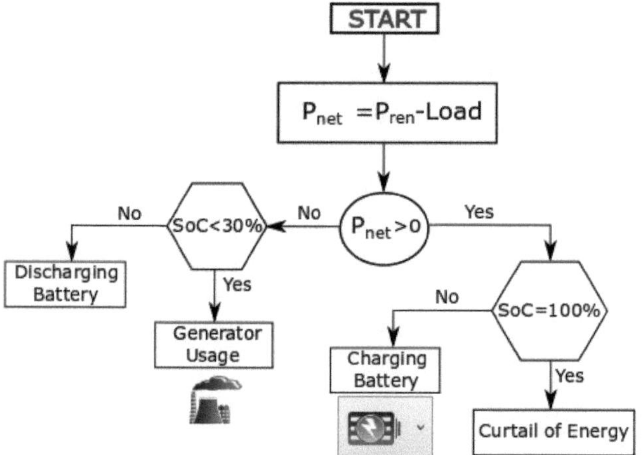

Figure 14: Smart Grid Algorithm Structure

The idea is that $\mathbf{P_{net}}$ is obtained by subtracting **Load** from $\mathbf{P_{ren}}$ (**Energy from Renewable Sources**). If $\mathbf{P_{net}}$ is positive, it is possible to store the energy in the battery until battery is fully charged. After battery is fully charged, excess is energy is called curtail of energy. Normally, curtail of energy is sold on spot, with relatively cheaper price to other parties.

If $\mathbf{P_{net}}$ is negative and the state of the charge of the battery is more than 30%, the battery is used to compensate the energy imbalance. However, if $\mathbf{P_{net}}$ is negative and the state of the charge of the battery is less than 30%, it is necessary to use generator to meet the unfulfilled energy demand.

6 Optimization of Smart Grid

The optimization is implemented using GA (Genetic Algorithm) which is part of Global Optimization Toolbox of Matlab. After cost of single unit of each component is plugged, the total cost is calculated considering the algorithm which is specified earlier. The total cost function is the **Objective Function** and the objective function variables are taken to be the number of components for Photovoltaic Panels, Wind Turbines and Batteries. These 3 variables are iteratively calculated and consequently total cost is obtained. The Genetic Algorithm is used to minimize the total cost by selecting the best combination of number of components for each variable.

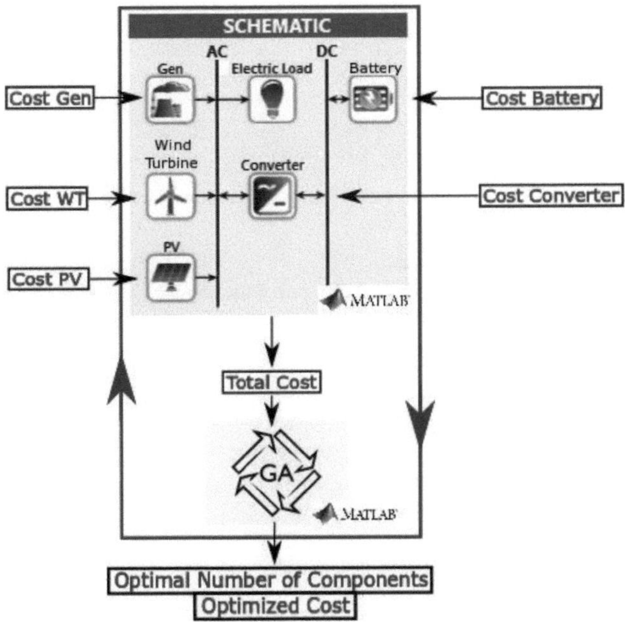

Figure 15: Optimization of Smart Grid using GA (Genetic Algorithm)

6.1 Optimization Algorithm

Optimization Algorithm consists of two parts:

1. Main Smart Grid Script file

2. Objective Function **Function** file

Main script file calls the objective function within GA algorithm and then optimized measurements are used to plot the main phenomena and to calculate the total CO2 emissions.

Listing 3: Main Smart Grid Script file

```matlab
clc; clear; close all
%% Generation of Hourly Load profile for a year with Matlab
global P_DG SoC Curt Ps COST_DG_OM COST_DG_TOTAL COST_TOTAL P_ren
    Pw nPs nPw nBat
load 'LOAD.csv'
LD=300*repmat(LOAD,1,365); % Replication of an Load array
Load=LD+0.3*rand(size(LD)).*LD-0.3*rand(size(LD)).*LD; %
    Randomizing the loads for the day
MLDR=mean(Load,1); % Mean load of each day
TD=1:8760; % Indexing hours of the year
DY=1:365;   % Indexing days of the year
figure ('Name', 'Load Profile for 300 houses')
plot(TD,Load(:),'r');
hold on
xlabel('Hour of Year');
ylabel('Load Profile for 300 houses [W]');
load 'TEMP.csv';
load 'WIND_S.csv';load 'SOLAR_R.csv'
Sw=repmat(WIND_S,1,24*31);
Sw=Sw';Sw=Sw(:);Sw(8761:end)=[];
Sw=Sw+0.3*rand(size(Sw)).*Sw-0.3*rand(size(Sw)).*Sw;
Rs=repmat(SOLAR_R,1,24*31);
Rs=Rs';Rs=1000*Rs(:)/24;Rs(8761:end)=[];
Rs=Rs+0.3*rand(size(Rs)).*Rs-0.3*rand(size(Rs)).*Rs;
syms X
options = optimoptions('ga');
options = optimoptions(options,'InitialPopulationRange', [0;10]);
options = optimoptions(options,'PopulationSize', 300);
options = optimoptions(options,'FunctionTolerance', 1e-300);
options = optimoptions(options,'ConstraintTolerance', 1e-300);
options = optimoptions(options,'SelectionFcn', {  @
    selectiontournament 4 });
options = optimoptions(options,'Display', 'iter');
options = optimoptions(options,'PlotFcn', {  @gaplotbestf @
    gaplotbestindiv });
LB=[0,0,0];UB=[300,20,100];
[X,fval,exitflag,output,population,score] = ...
    ga(@(X) Objective_Function(X,Sw,Rs,Load),3,[],[],[],[],LB,UB
        ,[],[1:3],options);
Total_DG_ENERGY=sum(P_DG(:));
CO2_factor=0.85/1000; % Emissions in kg/Wh
CO2_emissions=Total_DG_ENERGY*CO2_factor; % Total CO2 emissions
Total_CURT=sum(Curt(:));
Total_Ps=sum(Ps(:));
%% Plots
```

```
41  figure('Name','State of Charge');
42  plot(SoC,'k');
43  hold on
44  xlabel('Hour of Year');
45  ylabel('State of Charge [-]');
46  hold off
47  figure ('Name','Energy Generated by Diesel Generator')
48  plot(P_DG,'b');
49  hold on
50  xlabel('Hour of Year');
51  ylabel('Energy Generated by Diesel Generator [W]');
52  hold off
53  figure('Name','Curtail of Energy');
54  plot(Curt,'r');
55  hold on
56  xlabel('Hour of Year');
57  ylabel('Curtail of Energy [W]');
58  hold off
59  figure('Name','Renewable Energy from PV panels');
60  plot(Ps,'g');
61  hold on
62  xlabel('Hour of Year');
63  ylabel('Renewable Energy from PV panels [W]');
64  hold off
```

Listing 4: Objective Function Function file

```
1   function COST_TOTAL= Objective_Function (X,Sw,Rs,Load)
2   global P_DG SoC Curt Ps COST_DG_OM COST_DG_TOTAL COST_TOTAL P_ren
        Pw nPs nPw nBat
3   nPs=X(1); % Number of photovoltaic elements.
4   nPw=X(2); % Number of wind turbine units.
5   nBat=X(3); % Number of battery units.
6   %% Wind turbine
7   Sci = 4;Sco = 10.7;
8   Pr = 1800; % Rated capacity [W]
9   for i=1:8760
10      if Sw(i)<Sci
11          Pw(i)=0;
12      elseif Sw(i)>=Sci && Sw(i)<= Sco
13          Pw(i)=Pr*(Sw(i)-Sci)/(Sco-Sci);
14      elseif Sw(i)> Sco
15          Pw(i)=Pr;
16      end
17  end
18  COST_Wt=2200+20; % Capital+O/M costs
19  %% Photovoltaic Panel (Power)
```

```matlab
EtaPV=0.1515; % Solar panel 110W 12V Mono
Apv=1;
Ps=EtaPV*Apv*Rs; % generated power from single unit
COST_Spv=72.62; % cost of single module [$]   Solar panel 110W 12
    V Mono
%% Battery
SoC = zeros(8760,1);
SoC(1) = 0.8; % Initial SoC
SoC_min = 0.3; % Min. SoC
SoC_max = 1; % Max. SoC %
COST_Bat=410; % price of single battery unit [$]
CapBat=3110*nBat; % capacity of the battery [W]
%% Diesel Generator
P_DG = zeros(8760,1);
COST_DG_OM=0.326e-03; % price of single energy unit (O&M) of gen.
    power
COST_Gen=14400; % price of single unit generator
%% Energy management system
P_ren=nPs*Ps(:)+nPw*Pw(:);
Pnet=P_ren(:)-Load(:);
Curt = zeros(8760,1);
for h=2:8760
    if Pnet(h)>=0
        if SoC(h-1)<1
            SoC(h)=(Pnet(h)+SoC(h-1)*CapBat)/CapBat;
            if SoC(h)>1
            Curt(h)=SoC(h)*CapBat-CapBat;
            SoC(h)=1;
            end
        elseif SoC(h-1)>=1
            Curt(h)=Pnet(h);
            SoC(h)=1;
        end

    elseif Pnet(h)<0
        if SoC(h-1)> SoC_min
            SoC(h)=(SoC(h-1)*CapBat+Pnet(h))/CapBat;
            if SoC(h)< SoC_min
                P_DG(h) = abs(SoC(h)*CapBat-SoC_min*CapBat);
                SoC(h)= SoC_min;
            end
        elseif SoC(h-1)<= SoC_min
            P_DG(h) = -Pnet(h);
            SoC(h)=SoC_min;
        end
    end
```

```
64  end
65  Total_DG_ENERGY=sum(P_DG(:))
66  COST_DG_TOTAL=COST_DG_OM*Total_DG_ENERGY+COST_Gen;
67  COST_TOTAL=nPs*COST_Spv+nPw*COST_Wt+nBat*COST_Bat+COST_DG_TOTAL;
68  end
```

7   Optimization Results

The optimization results that are obtained are shown below:

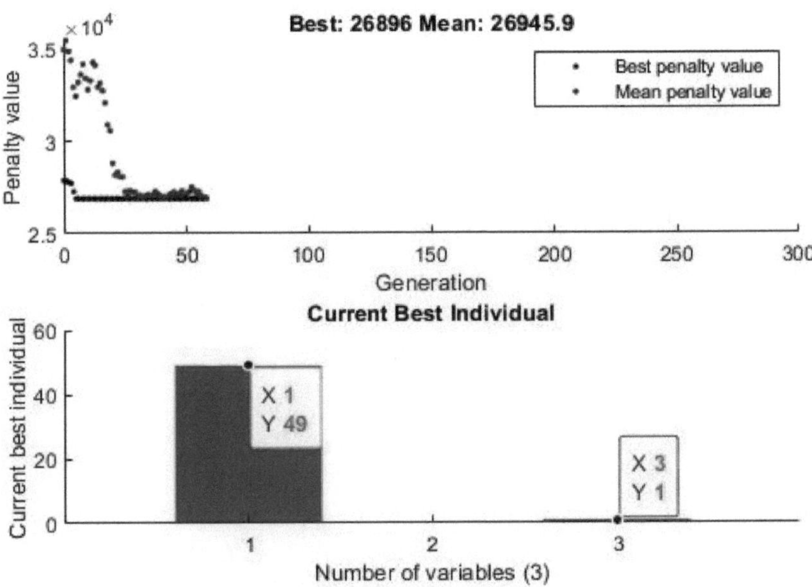

Figure 16: Optimized Results for Smart Grid Design

As it can be seen in Fig. 16, number of components optimized in terms of **quantity** are as following:

- Solar panel 110W 12V Mono - SuperWatt : **49 component(s)** (PV panels)

- AWS HC 1.8 kW : **0 component(s)** (Wind turbine)

- Discover 12VRE-3000TF: **1 component(s)** (Battery)

Total Cost is calculated to be:

$$COST_{total} = 2.689 \cdot 10^4 \ [\$] \tag{3}$$

8　Case Study Analysis

8.1　Energy Generated by Diesel Generator

As it is shown below, amount of energy generated by the diesel generator is much higher in the middle of the year which coincides with the lower amount of solar radiation. Since less energy is generated by the solar panels, the compensation is done using the diesel generator.
Total energy produced by the diesel generator is calculated:

$$\sum_{n=1}^{8760} P_{DG}(n) = 2.5939 \cdot 10^7 \ [W] \tag{4}$$

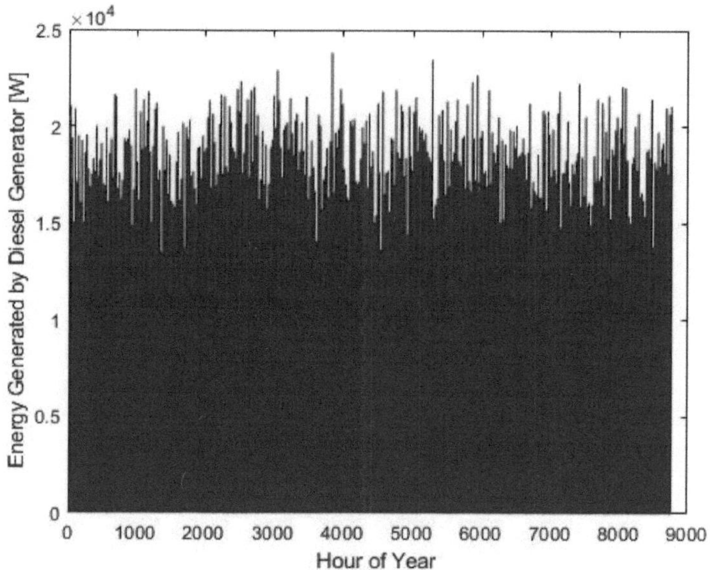

Figure 17: Energy Generated by Diesel Generator

8.2　CO2 Emissions

CO2 emissions generated by a diesel generator are estimated to be 0.8-0.93 kg CO2/kWh. Let's consider it as 0.85 kg CO2/kWh. Total CO2 emissions for a whole year can be calculated by multiplying the total energy generated by the diesel generator with the CO2 factor.

$$CO2_{total} = \sum_{n=1}^{8760} P_{DG}(n) \cdot CO2_{factor} \tag{5}$$

The equation is further simplified into:

$$CO2_{total} = \sum_{n=1}^{8760} P_{DG}(n) \cdot \frac{0.85}{1000} \tag{6}$$

Calculated CO2 emissions from Diesel Generator is shown below:

$$CO2_{total} = 2.2235 \cdot 10^4 \ [kg] \tag{7}$$

8.3 State of Charge

The state of charge throughout the year mainly depends on the solar radiation values since the only energy source for the battery are photovoltaic panels. As it can be observed, during the winter (middle of the year in Botswana), state of the charge of the battery drops significantly due to the shortage of solar radiation.

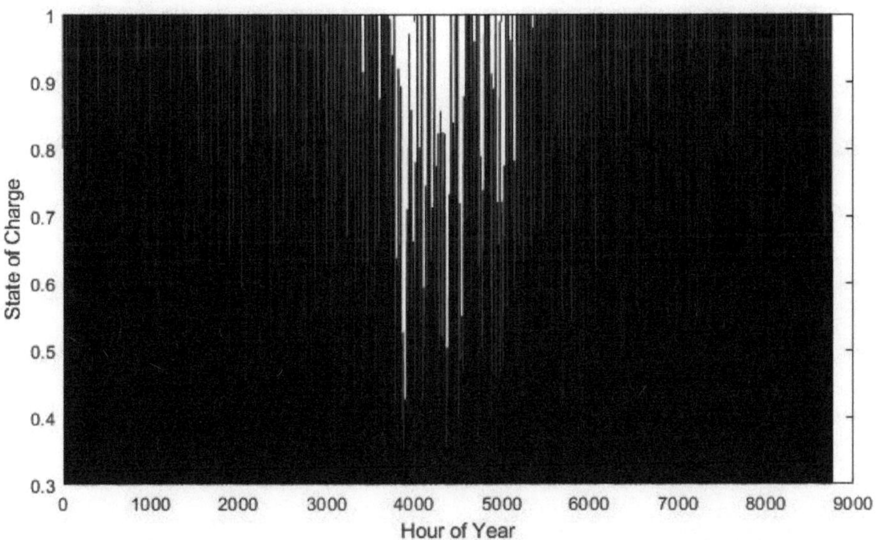

Figure 18: State of Charge

8.3.1 Curtail of Energy

As the state of the charge of the battery is at its full potential, the energy is transferred to the curtail of energy. This energy can normally be sold to other third party organizations.

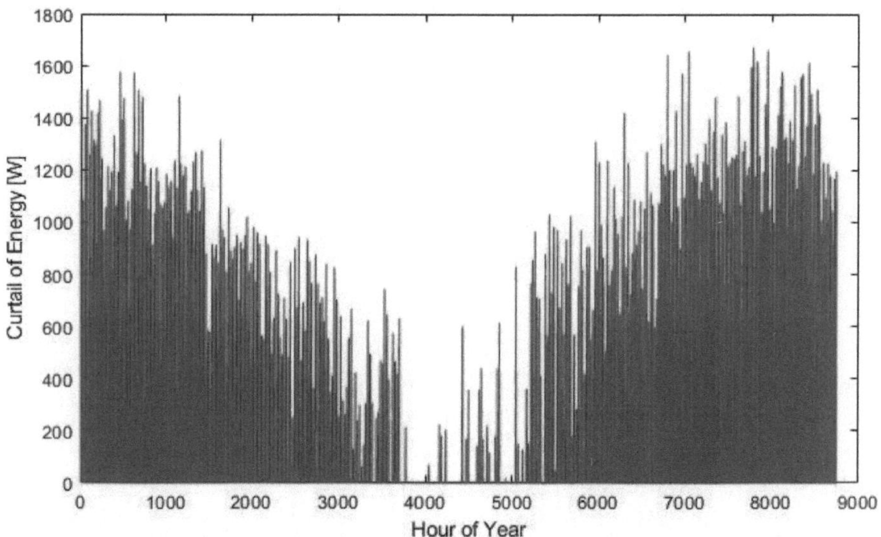

Figure 19: Curtail of Energy

The total energy as a form of curtail is calculated as below:

$$Curt_{total} = \sum_{n=1}^{8760} Curt(n) = 1.0609 \cdot 10^6 \ [W] \tag{8}$$

8.4 Renewable Energy from PV panels

The energy obtained from PV panels is illustrated below. The data matches with the solar radiation data which totally makes sense.

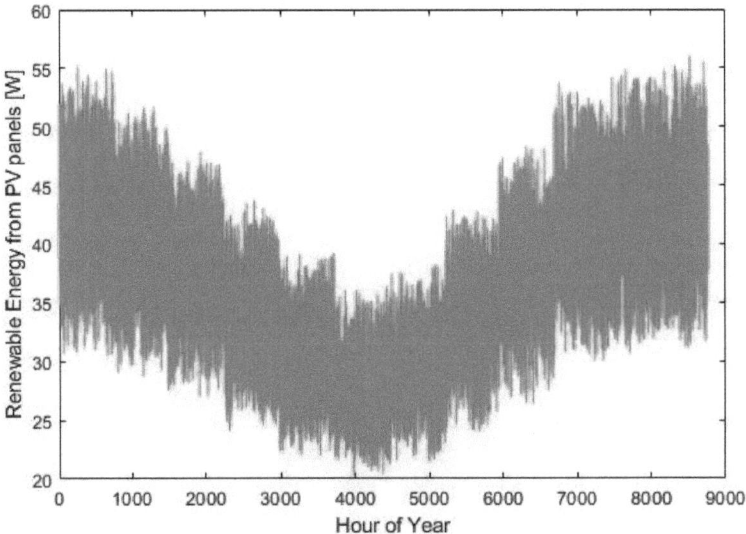

Figure 20: Renewable Energy from PV panels

The total energy as a form of renewable energy source from photovoltaic panels is calculated as below:

$$Ps_{total} = \sum_{n=1}^{8760} P_s(n) = 3.208 \cdot 10^5 \ [W] \tag{9}$$

9 Conclusion

In this practical work, Smart Grid Design and Optimization is managed through MATLAB. MATLAB is a powerful tool for numerical computations, specifically GA (Genetic Algorithm) is favored in this work for the optimization of the initially designed grid. Smart Grid Design is done pre-operatively using some useful data from HOMER Pro, including power curves for wind turbines as well as the efficiency data for photovoltaic panels etc..

As an objective of this practical work, the optimized design of Smart Grid and , consequently, optimized cost of the grid design is obtained through Global Optimization Algorithm.

In this practical work, 300 houses in rural region of Botswana are taken into account for the Smart Grid Design. Since the work doesn't really take into account of payback and other investment related measures, the design favors the energy from diesel generators which is not really case for more complex numerical tools such as HOMER Pro.

As a future work, the algorithm can be further improved and these economical measures can be taken into account for more complex solutions.